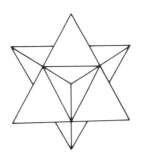

Cut & Assemble
3-D GEOMETRICAL SHAPES

10 Models in Full Color

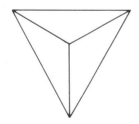

A. G. SMITH

Dover Publications, Inc., New York

Introduction and Instructions

The nature and structure of geometric solids has long fascinated mathematicians, scientists and philosophers. The Greek philosopher Plato, after whom the "Platonic solids" are named, and the German astronomer Johannes Kepler both made important contributions to the understanding of these forms. Artists and designers have found a source of inspiration in the intrinsic beauty and sense of order of geometric solids.

The geometric solids included in this volume are: the tetrahedron (1 piece, Plate 1), the octahedron (1 piece, Plate 1), the cube (1 piece, Plate 2), the trapezohedron (2 pieces, Plate 2), the dodecahedron (2 pieces, Plates 3 and 4), the icosahedron (2 pieces, Plates 5 and 6), interpenetrating tetrahedrons (5 pieces, Plates 5 and 6), the small stellated dodecahedron (13 pieces, Plates 12, 13 and 14), the great stellated dodecahedron (22 pieces, Plates 3, 4, 7, 8, 9, 10 and 11) and the great dodecahedron (16 pieces, Plates 15 and 16).

General Directions

The recommended tools for constructing the geometric solids are: 1) an X-ACTO knife with a #11 blade, 2) water-soluble white glue, such as Elmer's or Sobo, 3) a scoring tool, 4) a straightedge for scoring and 5) a burnishing tool for applying pressure to glued joints.

Study the cover photographs and assembly diagrams to see how the pieces fit together. Also read the step-by-step assembly instructions accompanying the diagrams. Note special instructions printed beside some pieces before cutting them out. Then carefully and accurately cut out all pieces needed for a given form. Try each piece for fit before applying glue. All tabs are marked with a letter of the alphabet indicating the sequence of assembly (A, B, C, etc.). Glue should be applied only to the tabs, never to the receiving surface.

For neater and more accurate results, score along all fold lines before folding.

It is advised to begin with the simple forms (tetrahedron, octahedron and cube) before moving on to the more complex forms. Patience and care will be rewarded with a pleasing result.

Of the assembly diagrams that follow, those for the first and easiest form, the tetrahedron, and for the last and hardest form, the great dodecahedron, are especially full and show the lettered tabs. The other diagrams show only the completed forms.

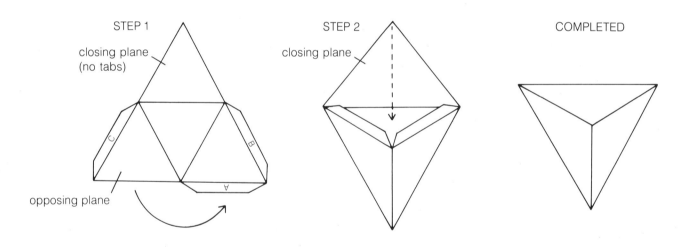

Assembling the Tetrahedron

1) Apply glue to tab A and attach the opposing plane.
2) Apply glue to tabs B and C and press down the closing plane.

Text continues after plates

Copyright © 1986 by A. G. Smith.
All rights reserved under Pan American and International Copyright Conventions.

Published in Canada by General Publishing Company, Ltd., 30 Lesmill Road, Don Mills, Toronto, Ontario.

Cut & Assemble 3-D Geometrical Shapes is a new work, first published by Dover Publications, Inc., in 1986.

International Standard Book Number: 0-486-25093-8

Manufactured in the United States of America
Dover Publications, Inc., 31 East 2nd Street, Mineola, N.Y. 11501

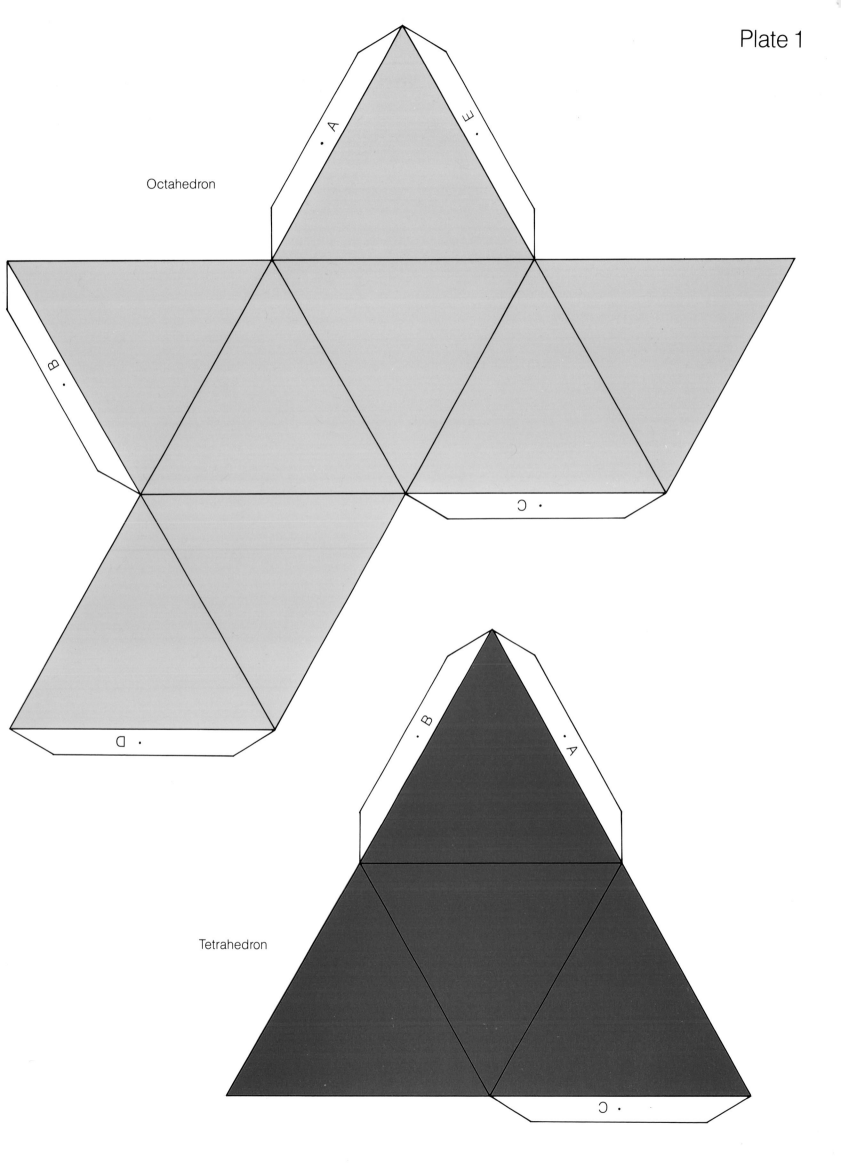

Plate 1

Octahedron

Tetrahedron

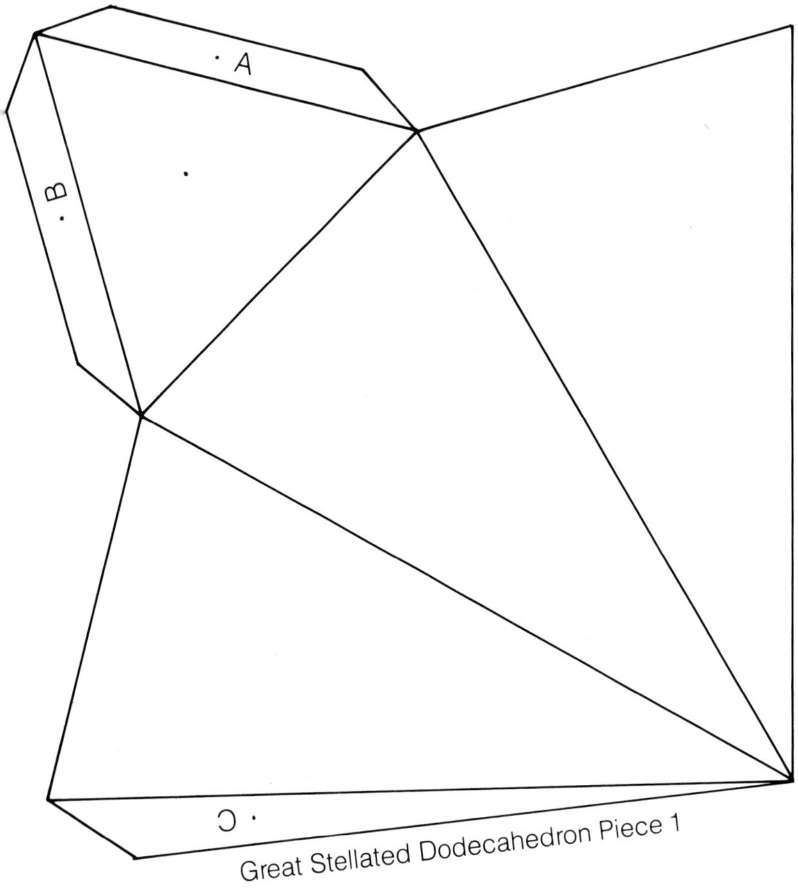

Plate 3

Great Stellated Dodecahedron Piece 2

Great Stellated Dodecahedron Piece 1

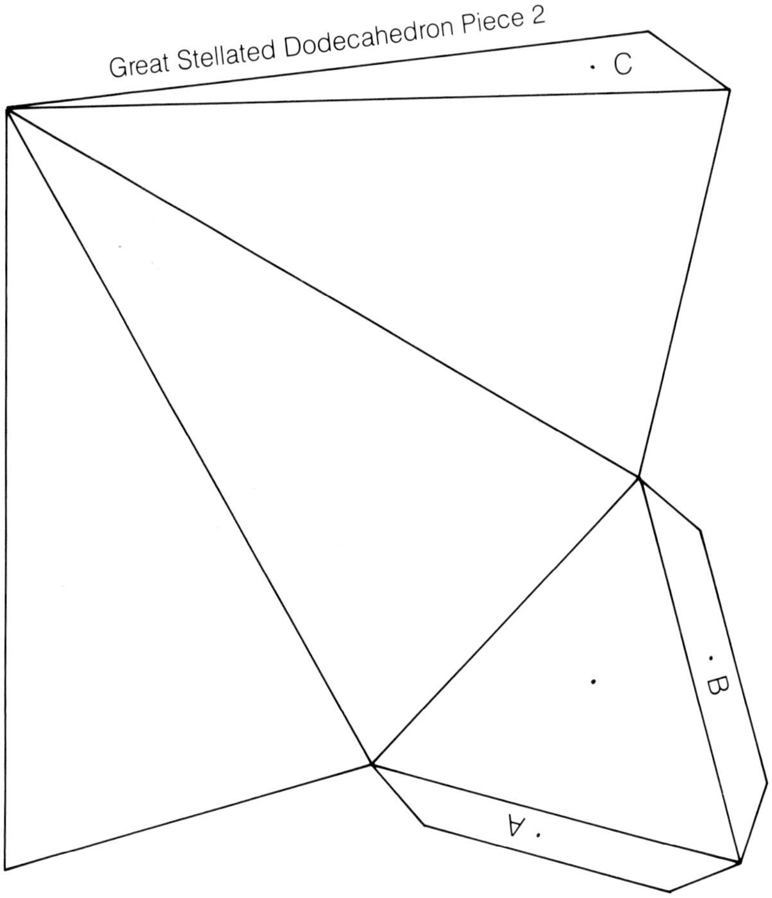

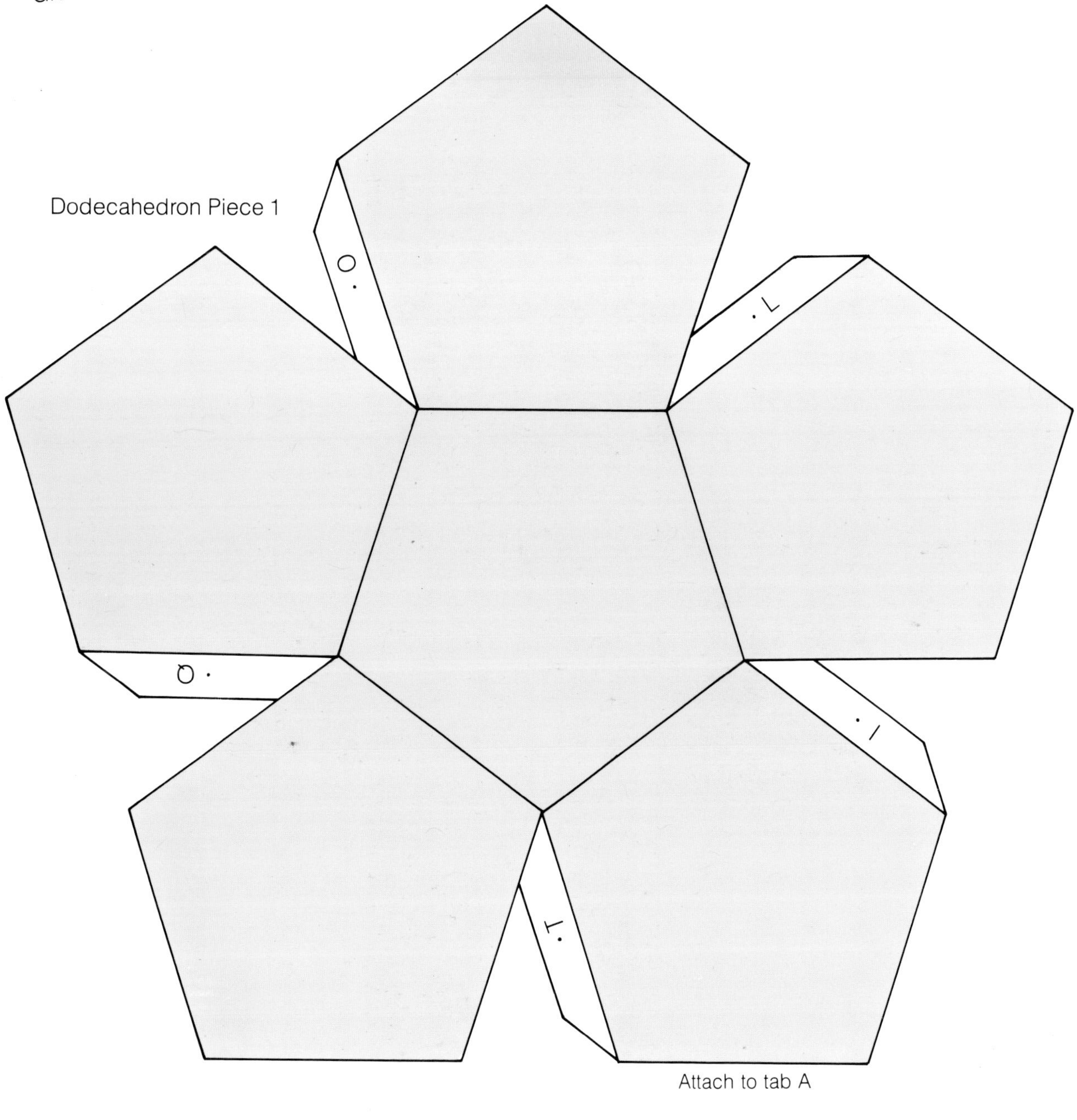

Dodecahedron Piece 1

Attach to tab A

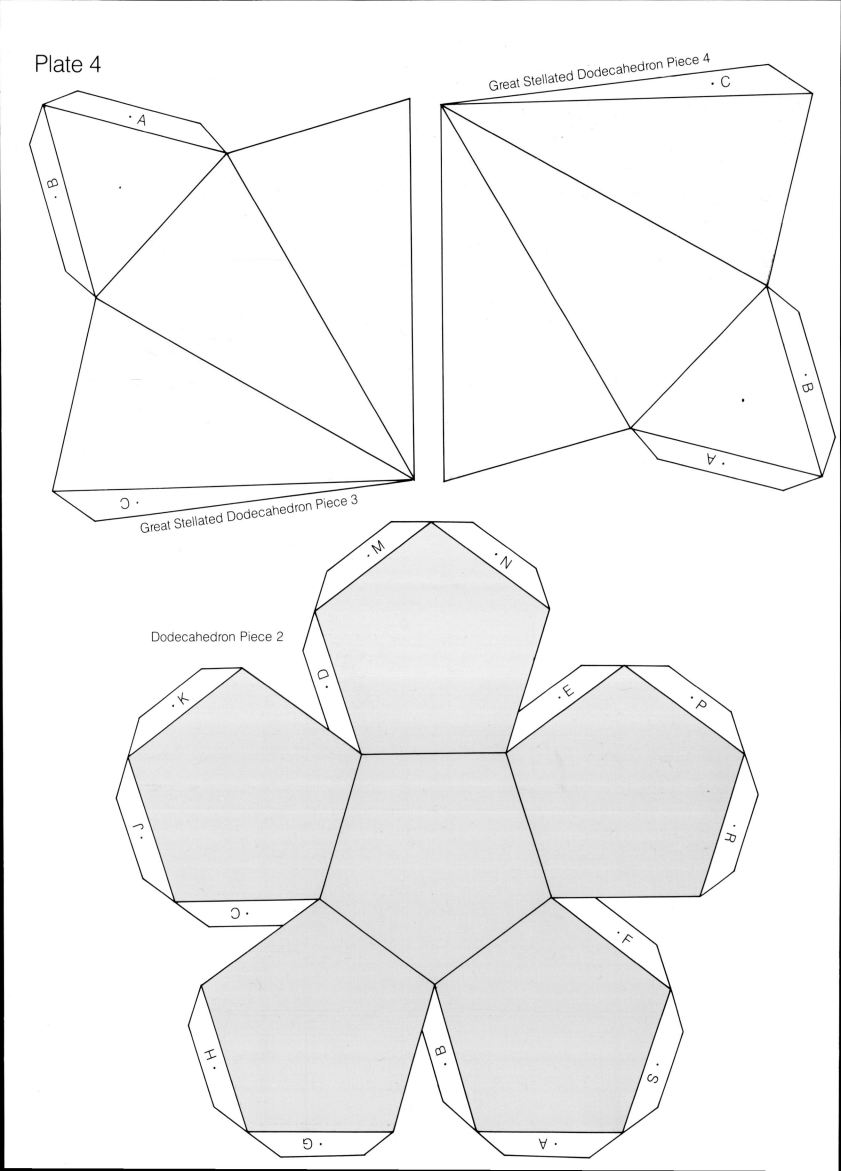

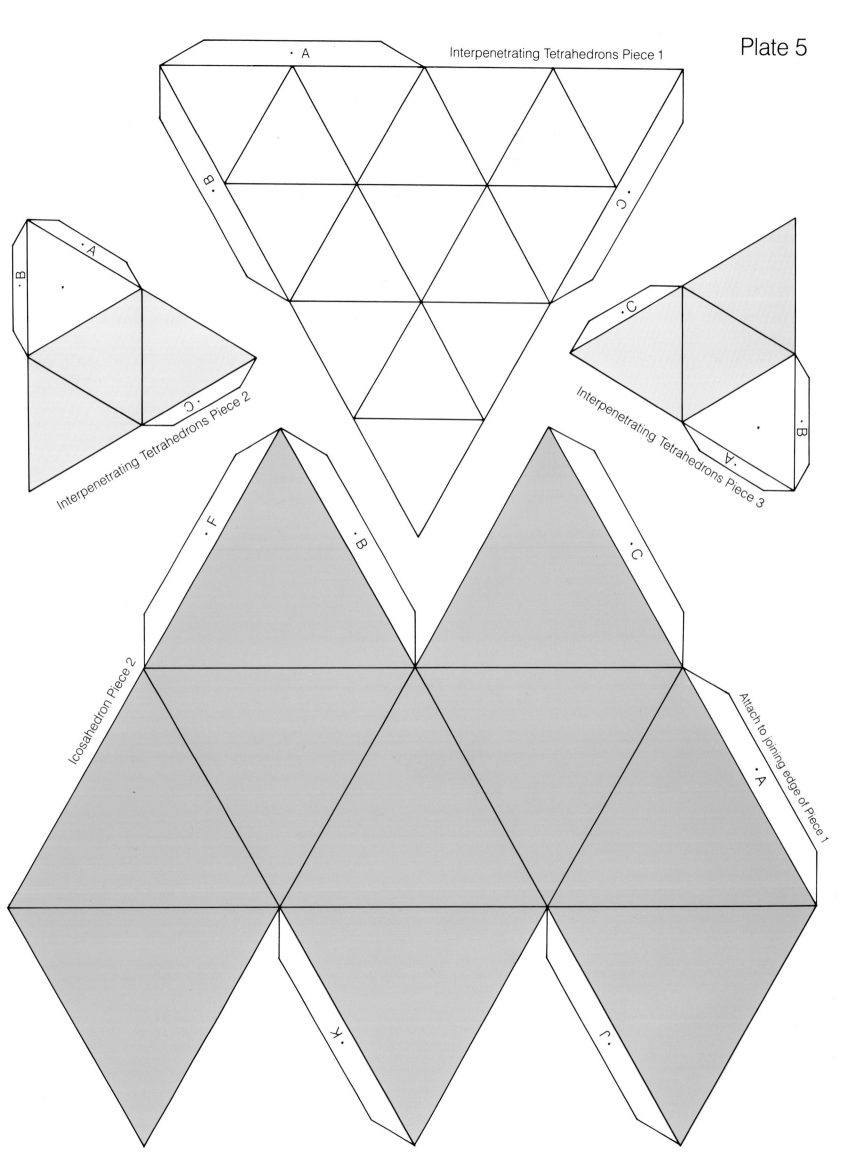

Great Stellated Dodecahedron Piece 6 (*second core piece*)

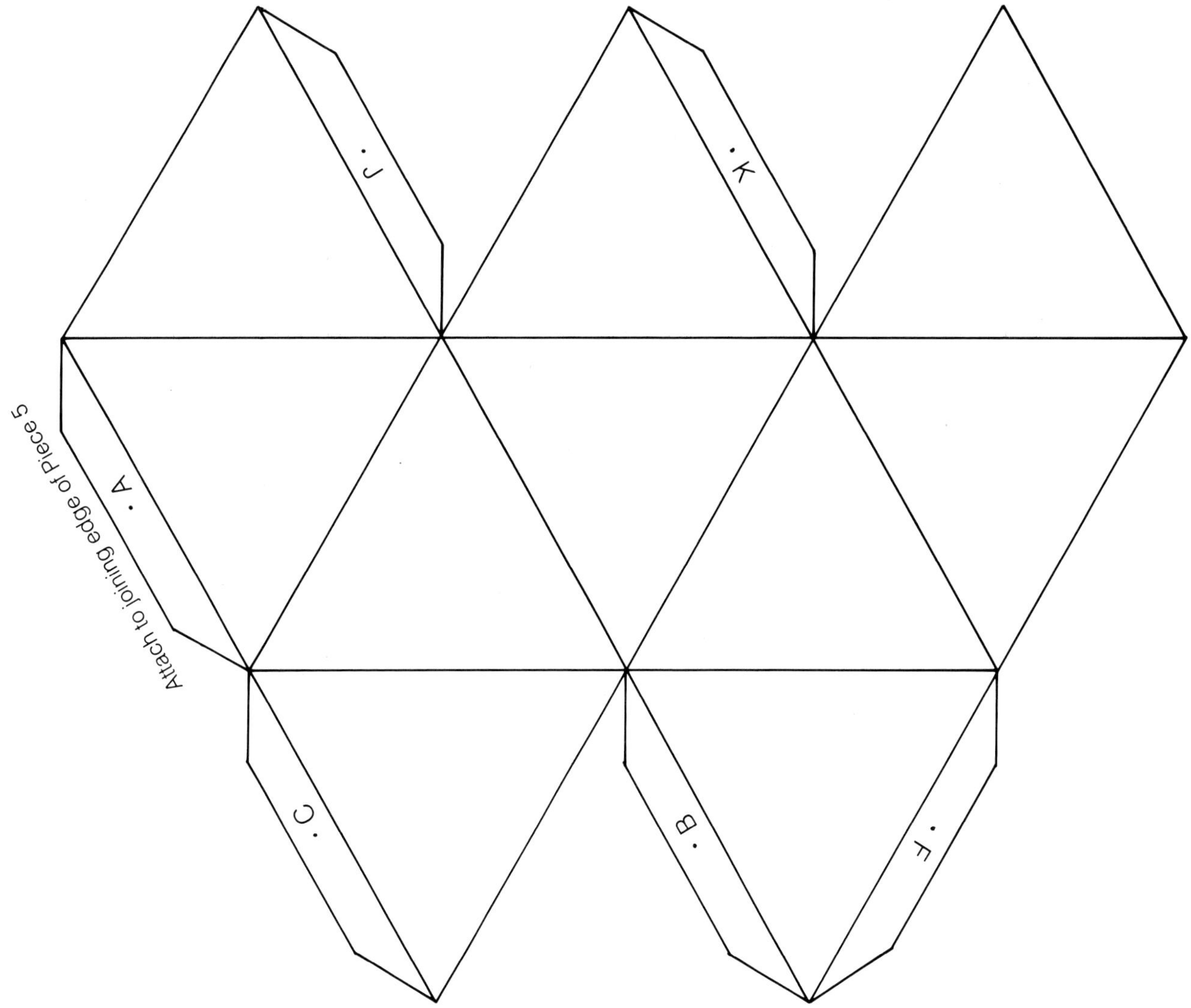

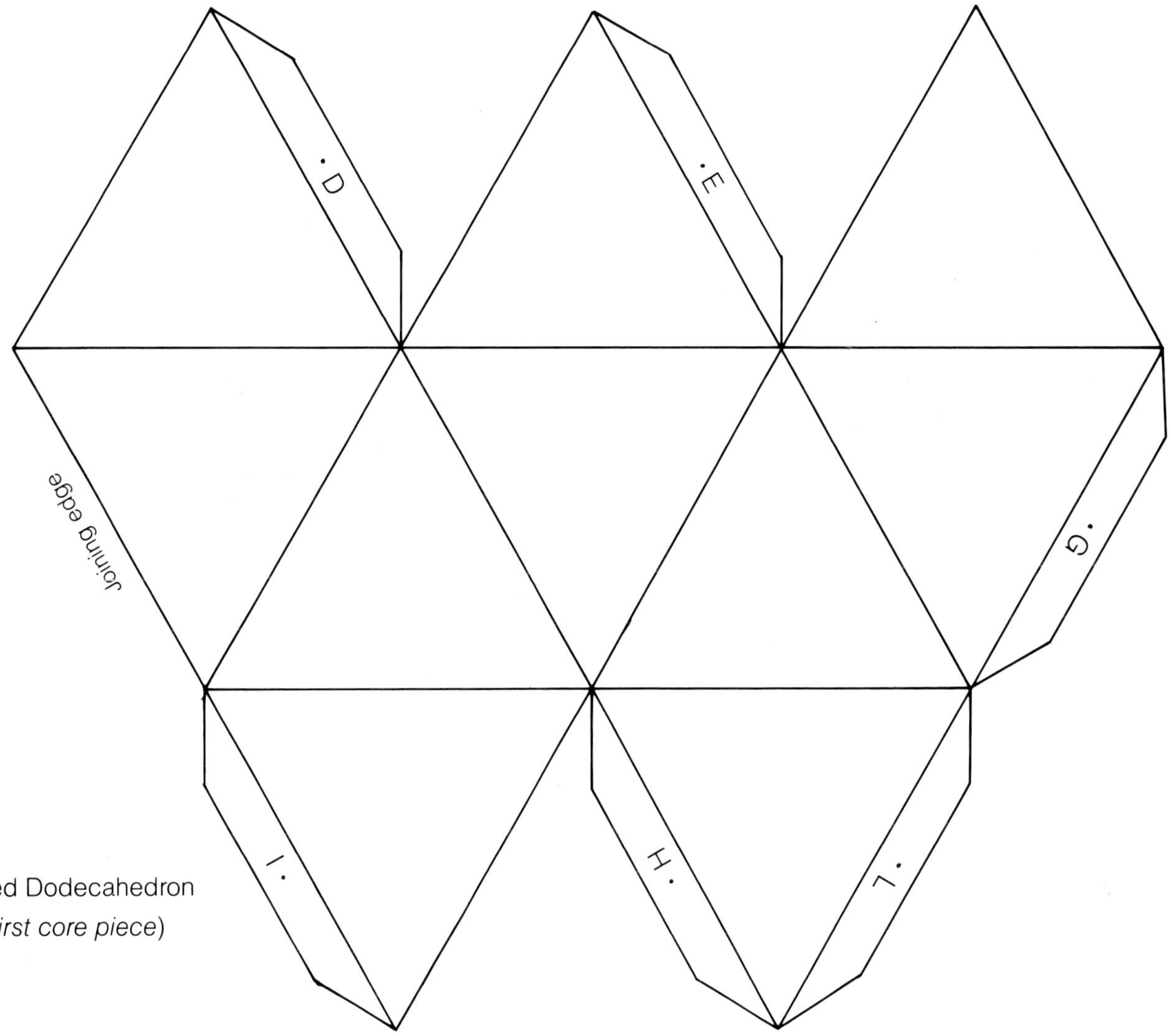

Plate 7

Great Stellated Dodecahedron
Piece 5 (*first core piece*)

Plate 10

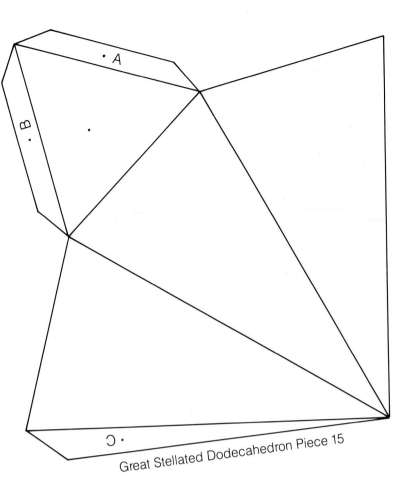
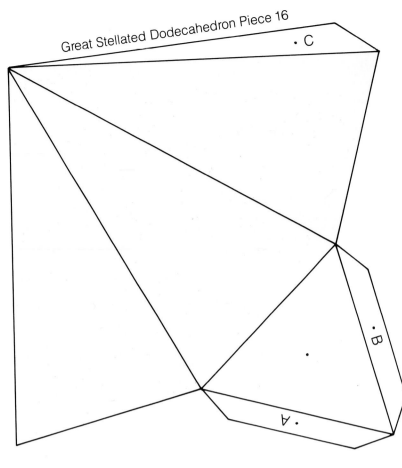
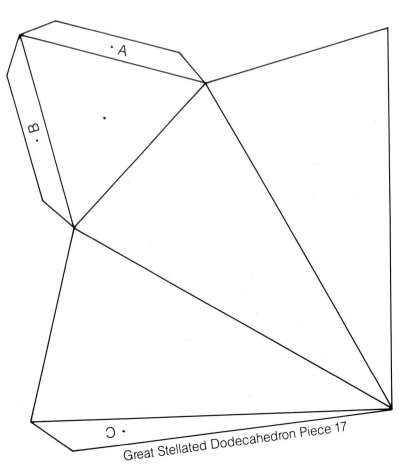
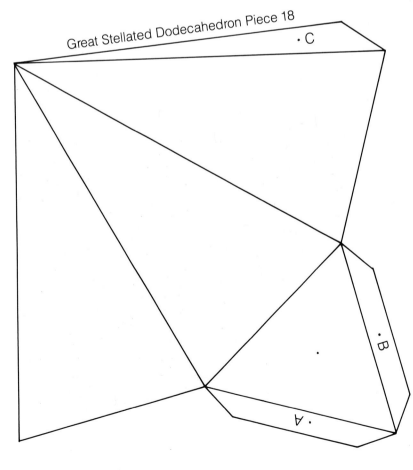

Plate 11

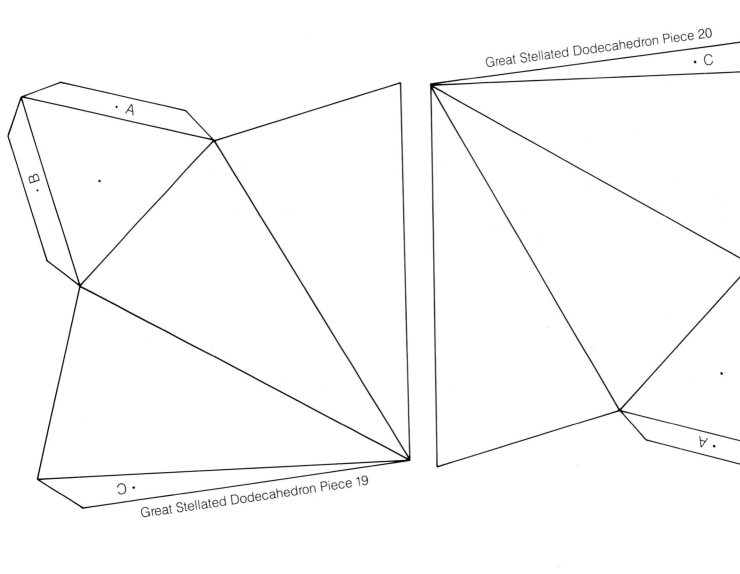

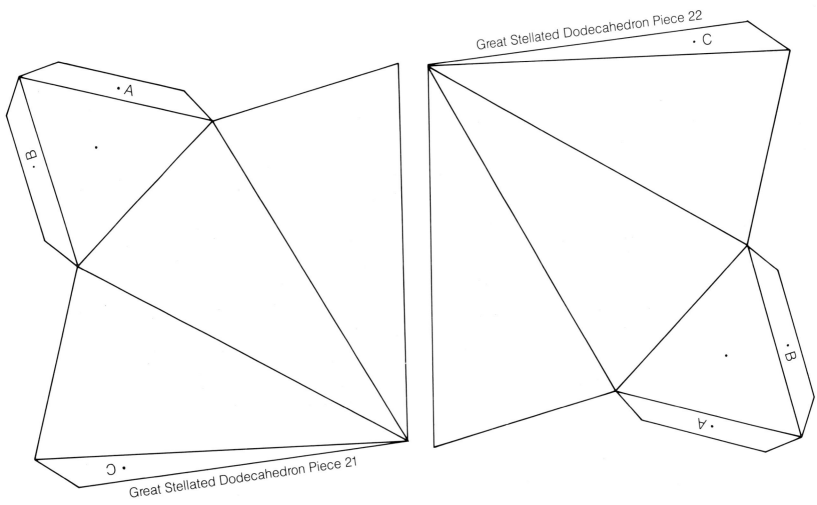

Plate 12

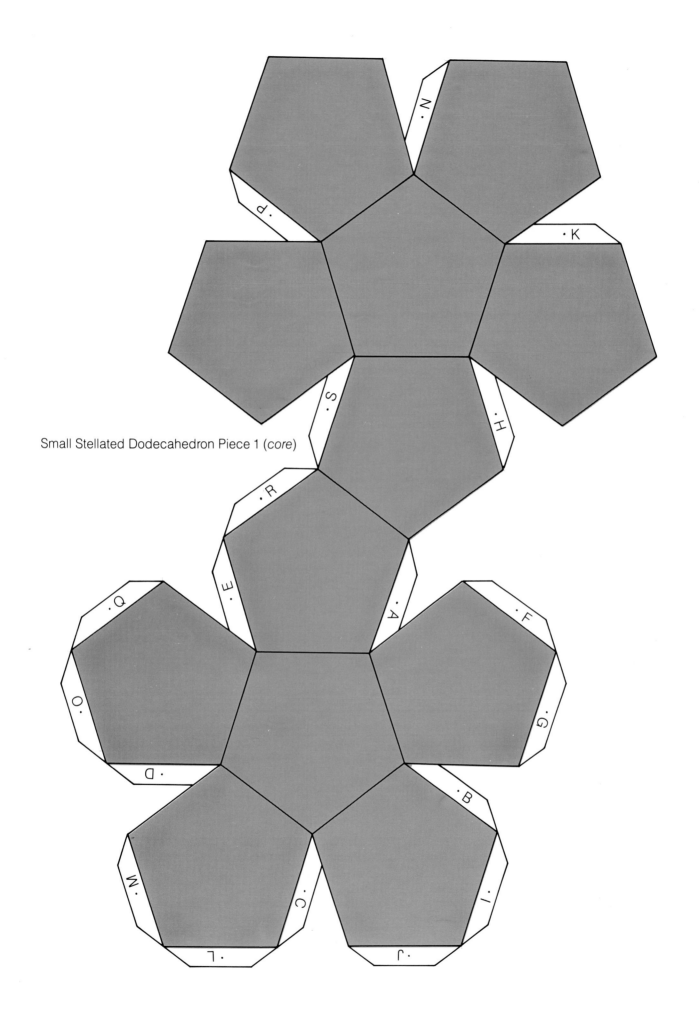

Small Stellated Dodecahedron Piece 1 (core)

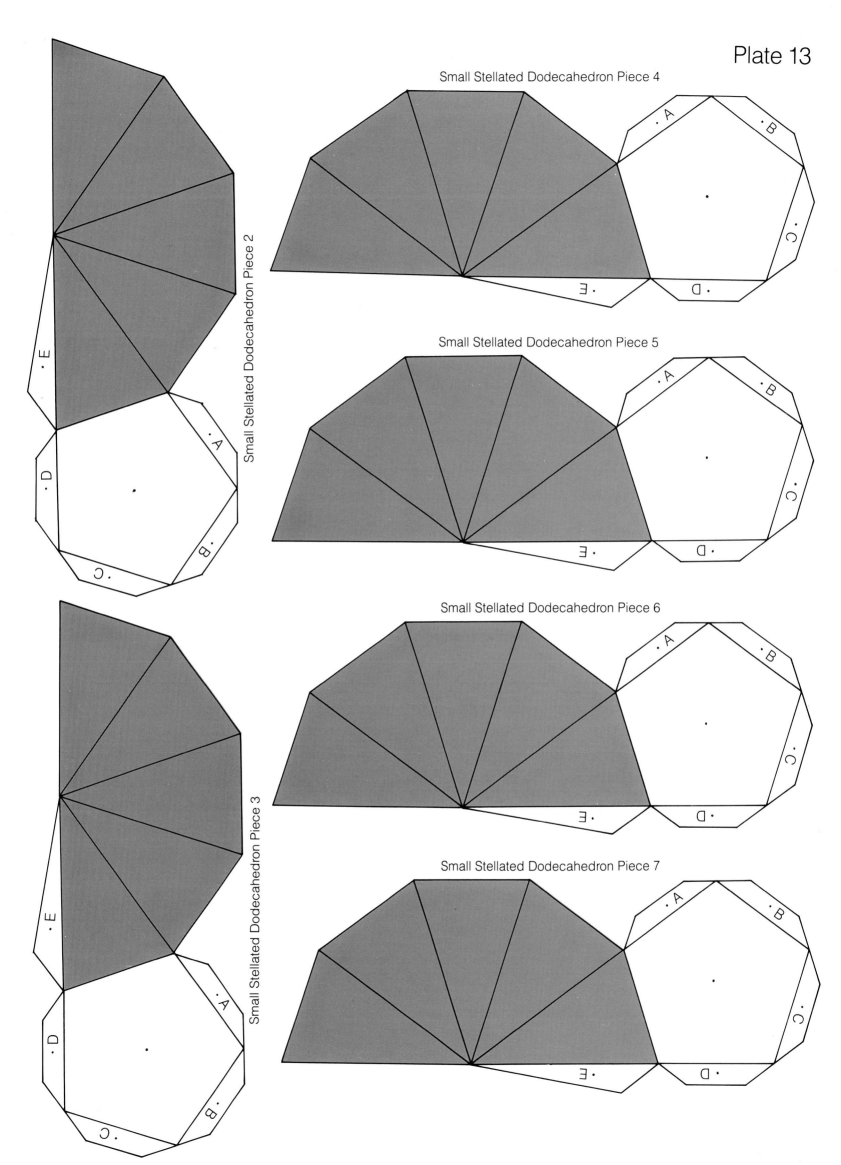

Plate 13

Small Stellated Dodecahedron Piece 2
Small Stellated Dodecahedron Piece 3
Small Stellated Dodecahedron Piece 4
Small Stellated Dodecahedron Piece 5
Small Stellated Dodecahedron Piece 6
Small Stellated Dodecahedron Piece 7

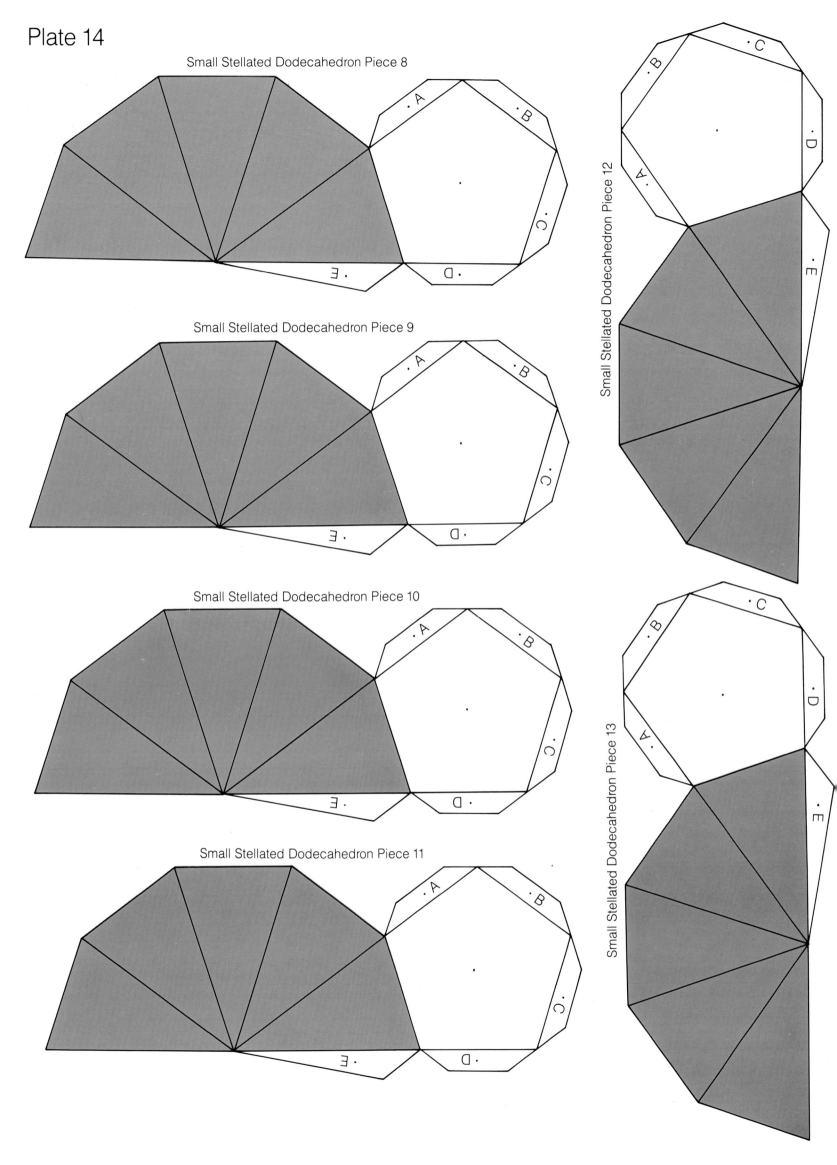

Plate 16

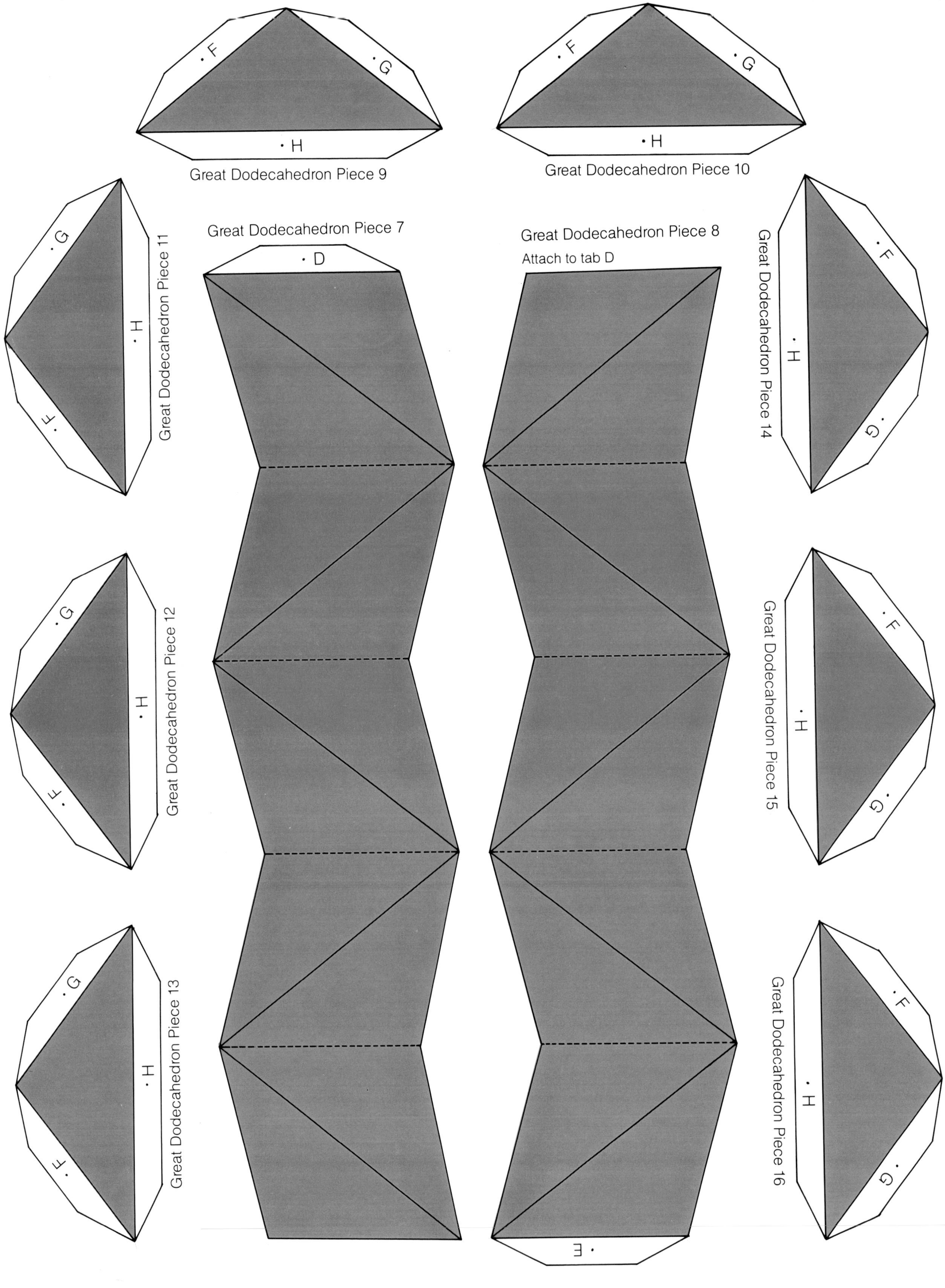
Great Dodecahedron Piece 9
• F
• G
• H
Great Dodecahedron Piece 10
• F
• G
• H
Great Dodecahedron Piece 7
• D
Great Dodecahedron Piece 8
Attach to tab D
• E
Great Dodecahedron Piece 11
Great Dodecahedron Piece 12
Great Dodecahedron Piece 13
Great Dodecahedron Piece 14
Great Dodecahedron Piece 15
Great Dodecahedron Piece 16

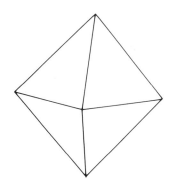

Assembling the Octahedron
1) Apply glue to tab A and attach the opposing plane.
2) Repeat Step 1 with tabs B and C.
3) Apply glue to tabs D and E and press down the closing plane.

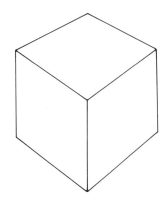

Assembling the Cube
1) Apply glue to tabs A and B and attach the opposing plane.
2) Repeat Step 1 with tabs C and D.
3) Apply glue to tabs E, F and G and bring down the closing plane.

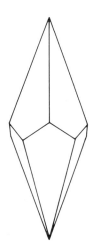

Assembling the Trapezohedron
1) Apply glue to tab A of Piece 2 and attach the indicated plane edge of Piece 1.
2) Apply glue to the long tab B and attach the opposing plane.
3) Apply glue to tab C and attach the opposing plane.
4) Repeat the tab-C operation in sequence (D, E, etc.).
5) Finish by gluing the closing plane down to tabs J, K and L.

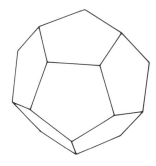

Assembling the Dodecahedron
1) Apply glue to tab A of Piece 2 and attach the indicated plane edge of Piece 1.
2) Assemble the Piece 2 half in the alphabetical sequence of the tabs (B, C, D, etc.).
3) Continue the process, bringing the closing plane down on tabs Q, R, S and T.

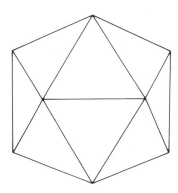

Assembling the Icosahedron
1) Apply glue to tab A of Piece 2 and attach the indicated plane edge of Piece 1.
2) Assemble the connected pieces in the alphabetical sequence of the tabs (B, C, D, etc.).
3) Continue the process, bringing the closing plane down on tabs K and L.

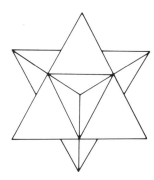

Assembling the Interpenetrating Tetrahedrons
1) Follow the instructions for the tetrahedron to assemble the large tetrahedron (Piece 1) and four small tetrahedrons (Pieces 2 through 5).
2) Apply glue to one side of each small tetrahedron and glue the sides to the positions indicated on the four sides of the large tetrahedron.

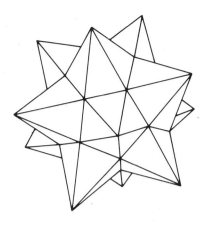

Assembling the Small Stellated Dodecahedron

1) Assemble the core (the small dodecahedron, Piece 1) by the alphabetical sequence of the tabs.
2) Assemble the 12 five-sided pyramidal forms (Pieces 2 through 13).
3) Apply glue to the bases of the five-sided forms and attach them to the dodecahedron.

Assembling the Great Stellated Dodecahedron

1) Assemble the core (the small icosahedron, Pieces 5 and 6 on Plate 7) by the alphabetical sequence of the tabs.
2) Assemble the 20 three-sided pyramidal forms (Pieces 1–4 and 7–22).
3) Apply glue to the bases of the three-sided pyramidal forms and attach them to the icosahedron.

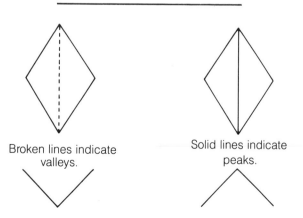

Broken lines indicate valleys.

Solid lines indicate peaks.

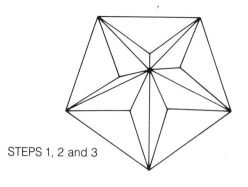

STEPS 1, 2 and 3

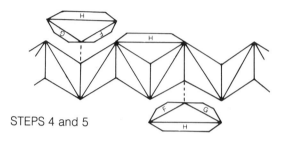

STEPS 4 and 5

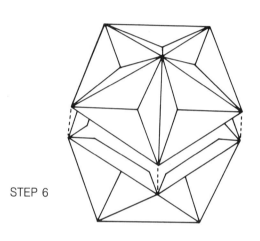

STEP 6

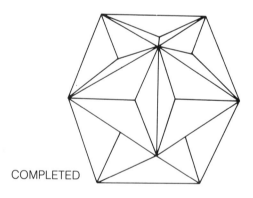

COMPLETED

Assembling the Great Dodecahedron

1) Cut out and score the two large pinwheel shapes (Pieces 1 and 2).
2) Apply glue to tabs A and B of the two inserts (Pieces 3 and 4) and attach them to the indicated edges of Pieces 1 and 2.
3) Glue the C tabs under their opposing planes.
4) Attach the two long bands (Pieces 7 and 8) at tabs D and E.
5) Attach the 10 triangular inserts (Pieces 5, 6 and 9–16) to the band at tabs F and G.
6) Apply glue to the tabs of the band marked H and attach the upper and lower completed star figures.